AERO SERIES VOLUME 27

CONVAIR
F-106

"Delta Dart"

By
William G. Holder

ISBN 0-8168-0600-4

AERO PUBLISHERS, INC.

329 West Aviation Road, Fallbrook, CA 92028

Dedication
To Beth

Library of Congress Cataloging in Publication Data

Holder, William G   1937-
    Convair F-106, "Delta Dart."
    (Aero series ; v. 27)
      1.   Delta Dart (Jet fighter plane)   I. Title.
UG1242.F5H63   358.4'3   75-15272
ISBN 0-8168-0600-4

FOREWORD

More than twenty years ago, while attending an F-102 Cockpit Configuration Meeting in San Diego, I noticed a fellow carrying a large block of fine wood across the hangar. When asked what he was going to do with it, he replied, "I'm going to help build a mock-up of a new fighter; it's to be called the F-102B!" The Old Girl has come a long way since those early days. Her designation was, of course, changed to "F-106" due to the extensive design changes which, among other things, allowed her to recapture the world's absolute speed record from the Soviets in 1959. Her responses to international crises such as the Pueblo incident and various airspace intrusions have been rapid, effective signals of national authority in every instance.

You won't find a "106" pilot who doesn't consider himself indeed fortunate to have flown the very best. Great credit is due Dick Johnson, Chief Convair Test Pilot, and his dedicated staff of talented engineers who worked so hard to make her what she is. My thanks also to Bill Holder for making it possible for readers to share a great experience.

JOE ROGERS
Colonel, USAF (Retired)
(Holder of World's single-engine
  speed record in 1959 in an F-106)

Col. Joe Rogers after his record-breaking F-106 speed run.   (USAF Photo)

3

ACKNOWLEDGEMENTS

1. General Dynamics Corporation/Mr. Fred DeFrance/Mr. Cliff Laechelin
2. AFLC Historical Archives
3. Aerospace Defense Command INTERCEPTOR MAGAZINE
4. Aeronautical Systems Division Office of Information/Mr. Bob Maltby
5. AFLC Office of Information Lt. Col. Nick Apple
6. Tyndall AFB Office of Information
7. Air Force Museum Research Department
8. Air National Guard Public Information Office
9. General Dynamics Corporation/Mr. Lou Garrett
10. Col. Joseph Rogers/Air Defense Command
11. Captain Don Carson/Secretary of Air Force Office of Information
12. L/C Jensen/Secretary of Air Force Office of Information
13. Hughes Aircraft Corporation, Public Affairs Office
14. Air Force Magazine
15. Dale Witt-Photography
16. Bob Shenberger-Photography
17. Mrs. Phyllisann Trimble-Typing

TABLE OF CONTENTS

the
F-106A
and
F-106B
delta dart

Cover of F-106 Manual (USAF Photo)

Chapter I

Genealogy

The "Cadillac" of the Fighters . . . "The finest-flying fighter aircraft ever built by the United States" . . . "The last fighter designed and optimized to perform a single mission." These are but a few superlative statements made about a superlative aircraft.

The pilots who have flown the F-106, to the man, love her. And her outstanding performance and super-sleek lines belie her 1950's heritage. But the F-106 is far from being outdated and is still far superior for the Aerospace Defense Command (ADC) mission. Through her many years of service, her airframe has remained structurally sound, which could well stretch her operational lifetime into the 1980's.

The Delta Dart never evolved beyond the single-seat F-106A and the two-seat F-106B. However, through the 1960's, and even into the early 1970's, radical changes were proposed for the "Six" which in the eyes of many would have made her competitive with the new advanced fighters of the 1970's. But even though the "Six" never evolved

from her initial configuration, she carries an impressive family tree from which she acquired her heritage.

The 106's roots go deep—actually sifting back into the late 1940's time period.

Her story began early in 1949 with the advent of the forerunner F-102 program. The F-102 grew out of efforts to overcome an armor deficiency resulting from the Russian's early detonation of an atomic bomb. The USAF did not have an interceptor able to meet the threat of Soviet jet bombers carrying the bomb. Postwar economy had prevented active pursuance of the 1945-1946 experimental interceptor projects. The Air Force in 1949 urgently required an interceptor subject to controlled operation from ground detecting systems. Its speed should approach or exceed Mach 2 and, in combat, it would have to operate up to and over 60,000 feet. The F-106 would eventually fulfill these requirements but it would be years before they would be realized.

Early in 1949, the USAF Senior Officers Board reached two decisions. First, the interceptor require-

Formation of F-106 Delta Darts fly past famous Mount Rushmore faces.   (USAF Photo)

F-102, 0-61416, which is currently on display at the Air Force Museum at Dayton, Ohio.
(Robert Shenberger Photo)

This photo of the Air Force Museum's F-102 clearly shows its wing fences and **droop** *on the leading edge of the wing.* *(Robert Shenberger Photo)*

8

ment would be given high priority and the program would be accelerated. Secondly, a completely new development would be undertaken, rather than an attempt to produce an all-weather interceptor version based on a general purpose fighter airplane. The approach would be that of an integrated weapon system development.

In December 1949, USAF invited proposals for solving the electronics and control system problems for the proposed aircraft. Then, Engineering Project MX-1179 was established and proposals were requested from industry in January 1950. From evaluation of the 13 proposals submitted, the USAF selected that of the Hughes Aircraft Company. Hughes was declared winner in July 1950 and given a research contract. At the same time, North American Aviation was directed to continue development of the Eyeball Radar Scanner, Inertial Autonavigator, and Radar Power Supply—electronic equipment items for the integrated MX-1179 Fire Control System.

The next step then was to obtain a vehicle for the MX-1179 system. The airplane required was designated the All-Weather Interceptor, 1954—Engineering Project MX-1554. It was to be ready for tactical use late in 1954 and was to have a first line life span in the 1955-1959 time period.

In July 1950, USAF initiated the competition for this development requesting proposals from 19 aircraft companies. Six of the companies replied by

January 1951. In July 1951, USAF declared three of the proposals winners—Consolidated-Vultee (Convair), Republic and Lockheed.

Several major decisions were then made during the last half of calendar year 1951. First, the Lockheed entry was scrapped because not enough research and development funds were available to conduct three separate research projects. Lockheed's proposal had been rated lowest; since the company had the F-104 interceptor program to keep it in business. At the same time the Republic proposal, the highly advanced F-103 airplane, was designated a long-range experimental development project. While very promising, the F-103 airplane required a very advanced engine configuration for its operation. The afterburner of its J-67 engine was to be used independently as a ramjet, by-passing air around the regular turbojet. As it would be impossible for the F-103 to be producible for the 1955-1959 period, this airplane was planned to be developed as a successor to the 1954 All-Weather Interceptor.

This left only the Convair proposal for the MX-1554. So in October 1951, USAF directed acceleration of the interceptor program. The Convair concept proposed a delta wing airframe, similar to that of its Delta-Wing XF-92A research aircraft with the Westinghouse YJ-67-W-1 engine, and the aforementioned Hughes MX-1179 Fire Control System. Its possible production was based on the availability of the J-67 by June 1954 and of the MX-1179 by June

Early F-102 during a landing. Note parachute deployment while nose is still airborne. (USAF Photo)

9

Intake location on the F-102. The intakes were moved *rearward* for the F-106.   (Robert Shenberger Photo)

The highly-unique North American F-107 was a fighter the same vintage as the F-106. At one time this aircraft, which exhibited a top-of-fuselage engine intake location, was considered a direct competitor to the 106. It, however, was never produced.
(Photo by author)

The Republic F-105 Thunderchief, built during the same time period as the F-106, was built in much greater numbers.   *(USAF Photo)*

The Republic XF-103 was to be an advanced fighter which was to follow the Convair 106 design.   (Republic Photo)

The great-granddaddy of the F-106, the XF-92A.   *(USAF Photo)*

1953.

Toward the end of 1951, it became increasingly evident that estimates as to availability for operational inventory of the MX-1554 were too optimistic. If the hoped-for new interceptor failed to materialize for the late 1950's time period, the USAF inventory would have, at best, only an improved F-86D which would not approach requirements. The F-106 was still far in the future.

A new assessment of the situation had to be made and an intermediate airplane between the current inventory and the ultimate MX-1554 had to be found. Accordingly, Headquarters USAF directed the Air Materiel Command to study and evaluate the possibilities of other airplane developments; namely, Republic's XF-91 and North American's Sabre 45 (later the F-100).

But finally the Air Force decided the route to go was to expand on Convair's interim proposal. The F-102 was therefore accepted without having to undergo a flight competition. The interim interceptor was designated the F-102A; the ultimate airplane—the MX-1554—was to be the F-102B which would ultimately evolve into the F-106. The F-106 would therefore gain improved performance through design improvements on the F-102A airplane.

Finally work began on the F-102, but problems surfaced quickly. USAF became uneasy about Con-

vair's predictions on drag—or the absence of drag in the F-102. Convair's facts and figures just did not agree with government estimates.

In March 1953, the Air Force called a conference to iron out differences of opinion and to plan a course of action. A plan was worked out for use if Convair proved to be wrong. Convair suggested that, if the airplane did not live up to performance expectations, it would camber the wing leading edges and modify the wing tips. NACA suggested that, in addition, the airplane should be reshaped on the basis of the Whitcomb "area rule" theory to achieve a more ideal body configuration.

The following F-102 "ideal body" changes were eventually effected in an effort to solve the airplane's deficiencies:

1. Cambered leading edges and reflex wing tips were added.
2. The wing was moved rearward and the vertical fin relocated.
3. The fuselage was extended 7 feet in length.
4. The fuselage was reeingineered into what became popularly known as the "coke bottle" or "Marilyn Monroe" shape.
5. Internal components of the airplane were shifted.

In December 1953, the F-102 prototype in the original configuration was first flown. It proved out as a Mach .98 airplane with a 48,000 feet ceiling. It

An interesting off-spring of the XF-92/F-102/F-106 family tree was the Navy XF2Y-1X Sea Dart. (U.S. Navy Photo)

was a far cry from the Mach 2, 60,000 foot altitude aircraft the Air Force had wanted.

A further redesign was definitely necessary. And these further refinements improved the 102's performance up to a maximum capability of Mach 1.5. The changes consisted of extending the nose four feet to provide a better fineness ratio. A new plexiglass canopy, providing better pilot visibility and improved air flow, was substituted for the old. The engine ducts were redesigned, an aft fuselage fairing was added to the tail to improve airflow, and further changes were made to the wing camber to approach an optimum design. Although these changes were for the F-102, the F-106 would also be a beneficiary.

Initially, the F-102A and the F-102B (F-106) were to be developed jointly, but the changes made to attain the new F-102A configuration irrevocably separated the F-102A from the F-106. The F-102A was now married to the J-57 engine—the J-67 planned for the F-106 could not now be interchanged. (Eventually the J-75 would be selected for the 106.) With a structural limitation of Mach 1.5 the F-102A airframe was eliminated as the airframe for the F-106 without major modification.

In late 1953, a decision had been reached to designate the first ten F-102's as YF-102's. These included the two prototypes—the first prototype crashed and was destroyed in November 1953—and 8 other airplanes built to the original configuration. The four airplanes built to the first major redesign configuration were also designated YF-102's. These four were used entirely for flight test purposes for testing the evolving F-106.

The F-102 that entered tactical inventory in June 1956, was larger and heavier than the airplane sought in 1951. Its fuselage length had increased considerably and its combat weight had escalated some 5,000 pounds from that originally estimated. Its engine showed an increase in thrust of only about 2,300 pounds.

Its performance, needless to say, did not live up to the hoped-for Mach 1.88 with a combat altitude of 56,500—actual figures were Mach 1.22 and combat altitude of 52,500 feet. This, however, was somewhat above the lowest acceptable minimum considered.

The F-106, which got its original designation in 1956, had greatly diverged in development from the F-102. The F-106 prototype first flew in December of 1956 from Edwards Air Force Base. (The two-seat F-106B flew some 18 months later.) Early in 1957, the "Six" approached the old 1949 requirements attaining an altitude of 57,000 feet and a top speed of Mach 1.9.

A formation of four F-102's in Greenland skies during 1968. (USAF Photo)

An F-102 rotates and starts gear retraction during a 1969 takeoff from Stewart AFB, New York.
(USAF Photo)

Two highly-painted F-102 Delta Daggers during unknown time frame on patrol over Alaska.
(USAF Photo)

F-102A Delta Daggers of the 327th Fighter Interceptor Squadron over California during 1957.   (USAF Photo)

Speed brakes of an F-102 are clearly detailed in this photo. This particular Delta Dagger carries the insignia of the Idaho Air National Guard. *(USAF Photo)*

A sign of the 1960's. An F-102 moves in for an inspection of a Soviet turbo-prop Bear Bomber.

One of the initial artist's concepts of the F-106.   *(Convair Photo)*

The test birds for the two Darts—the F-106 on the left and F-102 on the right. Note the rearward location of the intakes on the Delta Dart.
(Air Force Museum Photo)

The F-106A, S/N 6452, was the second Delta Dart to be built and participated in the test program. (USAF Photo)

The highly-shined surface of this test F-106 is quite evident by the reflection of the USAF stencil. (USAF Photo)

Carrying bright red paint for observation purposes, this F-106A (the 12th built) participated in the flight test program. Photo was taken at Edwards AFB in 1958.
(USAF Photo)

Convair F-106 flight test organization at Holloman Air Force Base (and test aircraft) in 1959.
(Convair Photo)

A beautiful paint job graces F-106A (60464) which served as one of the Holloman Test Aircraft.
(Convair Photo)

The first photo of the first two-seat F-106B. 46 of the B version were built by Convair. (USAF Photo)

In September 1956, Convair had proposed an F-102C to fill a possible gap between the end of the service life of the F-102A and the introduction of the F-106. The F-102C airframe would have contained all the improvements made on the F-102A. Its engine would be the advanced, higher thrust J-57-P-47. The fire control system would be the MG-14 designed to operate with the MB-1 rocket and Falcon missiles. But the Air Force did not accept this proposal, relying rather on the F-106 being ready for tactical inventory beginning in mid-1959. This proved to be the case as the first F-106 squadron became operational in May 1959. A total of 277 F-106A and 46 of the two-seat F-106B versions were built between 1956 and 1959 at the Convair San Diego facility.

In 1959, the F-106 demonstrated to the world what a hot machine it really was. On December 15th, (then Major) Joe Rogers set a single-engine world's speed record of 1525.95 miles per hour. The run was accomplished at 40,000 feet over Edwards Air Force Base, California. The record still stands today (1977). A year later, the "Six" demonstrated an amazing feat by flying some 2500 miles without refueling. —ADC was waiting, and the F-106 was ready!—

F-106 PRODUCTION DATA

CONVAIR F-106A

56-451/452	(2)	F-106A
56-453/455	(3)	F-106A
56-456/462	(7)	F-106A
56-463/467	(5)	F-106A
57-229/239	(11)	F-106A
57-240/242	(3)	F-106A
57-243/245	(3)	F-106A
57-246	(1)	F-106A-64
57-2453/2455	(3)	F-106A-64
57-2456/2460	(5)	F-106A-70
57-2461/2465	(5)	F-106A-75
57-2466/2477	(12)	F-106A-80
57-2478/2485	(8)	F-106A-85
57-2486/2506	(21)	F-106A-90
58-759/771	(13)	F-106A-95
58-772/798	(27)	F-106A-100
59-001/030	(30)	F-106A-105
59-031/059	(29)	F-106A-110
59-060/086	(27)	F-106A-120
59-087/111	(25)	F-106A-125
59-112/135	(24)	F-106A-130
59-136/148	(13)	F-106A-135

TOTAL F-106A (277)

CONVAIR F-106B

57-2507/2513	(7)	F-106B
57-2514/2547	(34)	F-106B
58-900/904	(5)	F-106B

TOTAL F-106B (46)

TOTAL OF BOTH F-106A & F-106B 323

Note:
    57-2507 is a F-106B-10-CO

26

World's Fastest—Major Joseph W. Rogers, jacket in hand, prepares to board the U.S. Air Force F-106 all-weather jet interceptor which carried him to a new world's speed record. The veteran Air Defense Command pilot flew the F-106, built by Convair Division of General Dynamics Corporation, over an 18 kilometer course at Edwards Air Force Base, California.
(General Dynamics Photo)

The smiling gentleman on the right is Air Defense Command's Major Joseph W. Rogers, and his exuberance stemmed from a 1525.95 mph world record speed run in a Convair F-106 Delta Dart. Post flight performance analysis began immediately after the 35-year-old pilot landed. Exchanging information are (l to r) Colonel Royal N. Baker, Director of Flight Test at this Mojave Desert Air Force Flight Test Center; Convair engineering staff specialist, Richard L. Johnson, who flew initial tests in the F-106 nearly three years ago; Convair acting chief engineering test pilot, Chuck Myers (with back to camera); and Major Rogers.

(Convair Photo)

Chapter II

The Nature of the Bird

From the outside, the F-106 looks just an awful lot like the F-102. But on closer examination, the differences between the two aircraft become more pronounced. The F-106 took the lessons that were learned in the 102 program, blended them in with advanced 1950's technology, and came out with an aircraft which is far from being obsolete even by 1970's standards. Just sitting there the "Six" looks like it's doing about Mach 2!

The two birds are, of course, both delta-wingers with identical wing spans of 38 feet, 1.6 inches. But the F-106 Delta Dart is 70 feet, 8.8 inches long—some two feet longer than the F-102. Also, the "Six" is 20 feet, 3.3 inches high from the ground to the top of its vertical fin, about one foot less than the 102. The "Six" grosses out at some 18 tons and has an empty weight of slightly over 23,-000 pounds. The "Six's" gross weight is some 8,-000 pounds greater than its older 102 sister, and only 4,000 pounds less than the new F-15 Eagle.

The "Six's" clipped vertical fin is a noticeable difference between the two aircraft. The 106 tail has a much longer base where it joins the fuselage, and in order to have maintained the same leading edge angle, it would have made the tail too high. No F-106 flight stability is lost, however, since the "Six" has the same-sized tail area-wise (some 695 square feet) as the F-102. From the beginning the "Six" was also designed so as to fit into F-102 hangars.

The "Six" is powered by a Pratt and Whitney J-75 engine equipped with an afterburner. This powerplant develops some 50 percent more thrust than the smaller J-57 of the F-102. Externally the 106 engine's air intakes are located far back of the cockpit at the wing root, accentuating the wasp waist of the "Six's" "Coke Bottle" fuselage. The rearward movement of the intakes (from the F-102 location) provided an aerodynamic improvement for the fuselage design.

3/4 view of a "Six" preparing for takeoff on an ADC exercise at Kingsley AFB, Oregon. (USAF Photo)

airplane general arrangement

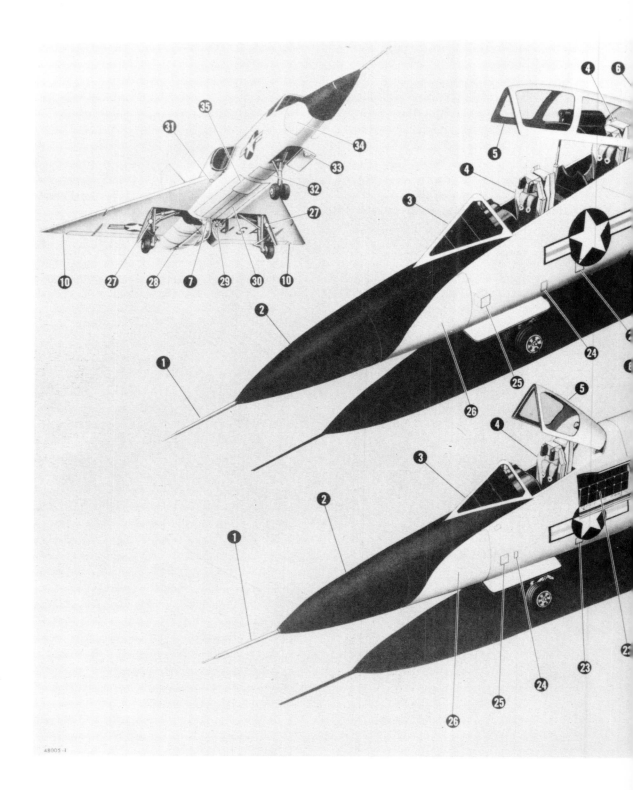

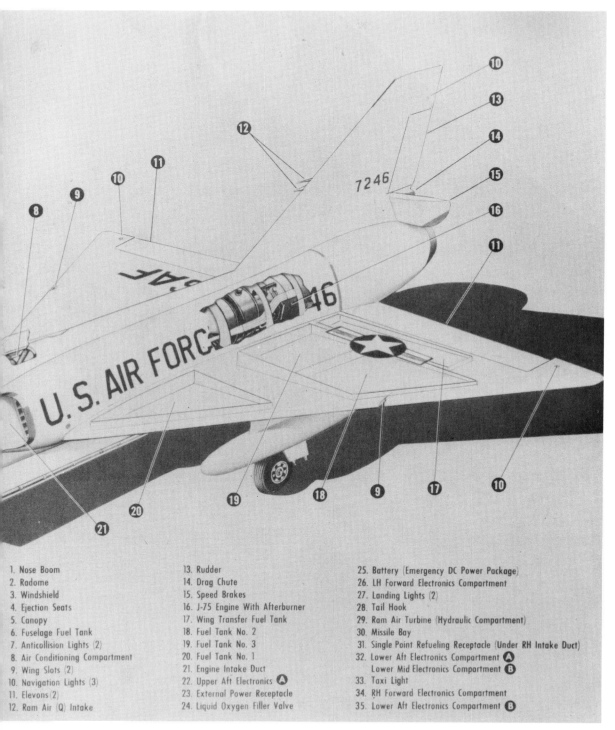

1. Nose Boom
2. Radome
3. Windshield
4. Ejection Seats
5. Canopy
6. Fuselage Fuel Tank
7. Anticollision Lights (2)
8. Air Conditioning Compartment
9. Wing Slots (2)
10. Navigation Lights (3)
11. Elevons (2)
12. Ram Air (Q) Intake
13. Rudder
14. Drag Chute
15. Speed Brakes
16. J-75 Engine With Afterburner
17. Wing Transfer Fuel Tank
18. Fuel Tank No. 2
19. Fuel Tank No. 3
20. Fuel Tank No. 1
21. Engine Intake Duct
22. Upper Aft Electronics Ⓐ
23. External Power Receptacle
24. Liquid Oxygen Filler Valve
25. Battery (Emergency DC Power Package)
26. LH Forward Electronics Compartment
27. Landing Lights (2)
28. Tail Hook
29. Ram Air Turbine (Hydraulic Compartment)
30. Missile Bay
31. Single Point Refueling Receptacle (Under RH Intake Duct)
32. Lower Aft Electronics Compartment Ⓐ
    Lower Mid Electronics Compartment Ⓑ
33. Taxi Light
34. RH Forward Electronics Compartment
35. Lower Aft Electronics Compartment Ⓑ

The major details of the F-106A and F-106B are shown.   *(USAF Photo)*

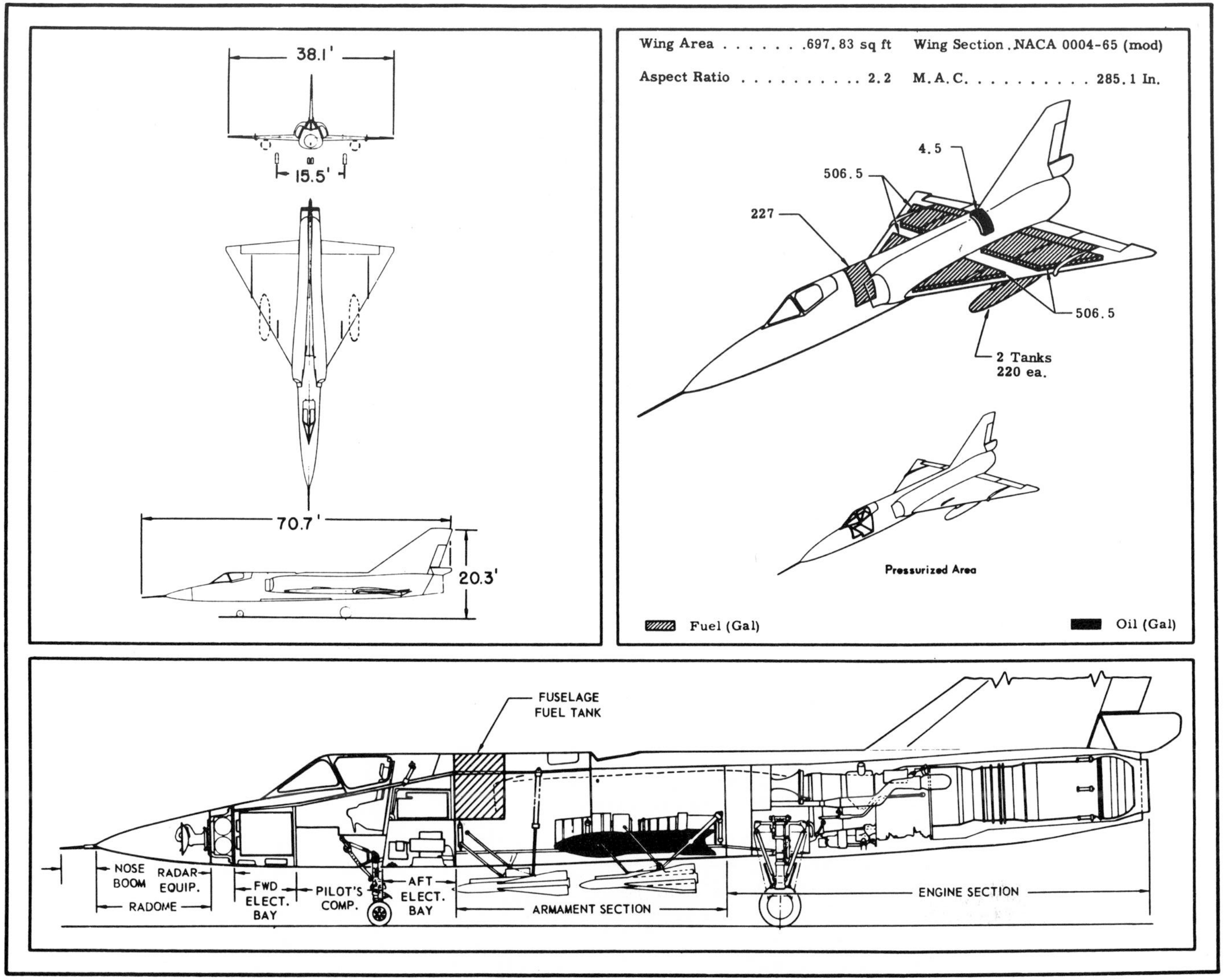

Wing Area . . . . . . 697.83 sq ft   Wing Section . NACA 0004-65 (mod)

Aspect Ratio . . . . 2.2   M.A.C. . . . . 285.1 In.

506.5

4.5

506.5

227

2 Tanks
220 ea.

Pressurized Area

■ Oil (Gal)

▨ Fuel (Gal)

38.1'

15.5'

70.7'

20.3'

FUSELAGE
FUEL TANK

ENGINE SECTION

ARMAMENT SECTION

AFT
ELECT.
BAY

PILOT'S
COMP.

FWD
ELECT.
BAY

RADAR-
EQUIP.

NOSE
BOOM

RADOME

Early USAF drawing showing internals of the F-106A.

(USAF Photo)

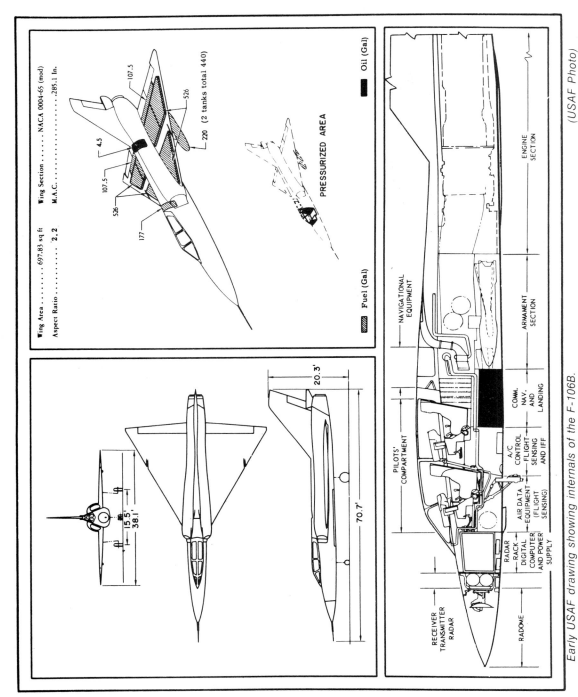

Wing Area . . . . 697.83 sq ft

Aspect Ratio 2.2

Wing Section . . . . . . NACA 0004-65 (mod)

M.A.C. . . . . .285.1 In.

107.5

526

220 (2 tanks total 440)

4.5

107.5

526

177

PRESSURIZED AREA

Fuel (Gal)

Oil (Gal)

NAVIGATIONAL EQUIPMENT

PILOTS' COMPARTMENT

A/C CONTROL FLIGHT SENSING AND IFF

AIR DATA EQUIPMENT (FLIGHT SENSING)

RADAR RACK DIGITAL COMPUTER AND POWER SUPPLY

RECEIVER TRANSMITTER RADAR

COMM. NAV. AND LANDING

ARMAMENT SECTION

ENGINE SECTION

RADOME

20.3'

70.7'

38.1'

15.5'

Early USAF drawing showing internals of the F-106B.

(USAF Photo)

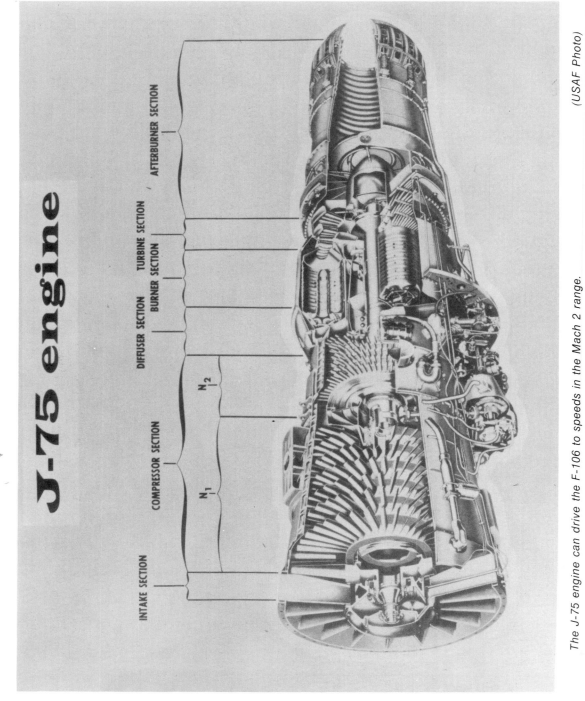

J-75 engine

INTAKE SECTION · COMPRESSOR SECTION · N_1 · N_2 · DIFFUSER SECTION · BURNER SECTION · TURBINE SECTION · AFTERBURNER SECTION

The J-75 engine can drive the F-106 to speeds in the Mach 2 range.

(USAF Photo)

A rear angle shot from the tail-pipe looking forward. Note the dual wheels of the front gear. (Dale Witt Photo)

fuel quantity data table

DATE: 10 DECEMBER 1960
DATA BASIS: ACTUAL

U.S. GALLONS AND POUNDS

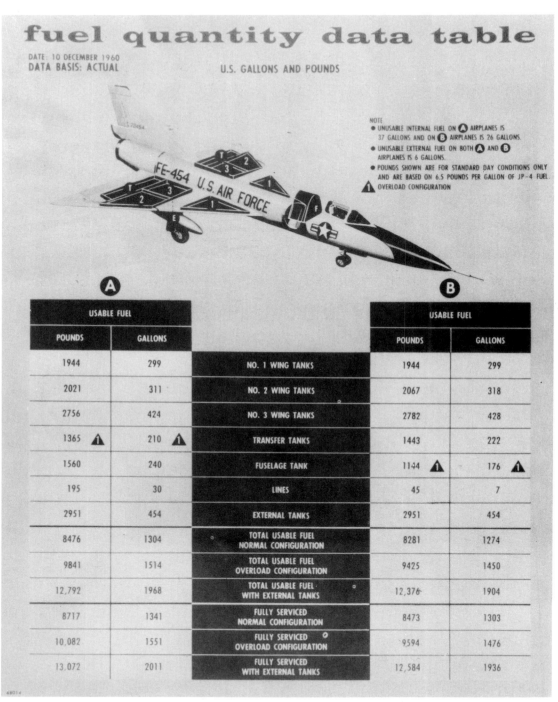

NOTE
- UNUSABLE INTERNAL FUEL ON (A) AIRPLANES IS 37 GALLONS AND ON (B) AIRPLANES IS 26 GALLONS.
- UNUSABLE EXTERNAL FUEL ON BOTH (A) AND (B) AIRPLANES IS 6 GALLONS.
- POUNDS SHOWN ARE FOR STANDARD DAY CONDITIONS ONLY AND ARE BASED ON 6.5 POUNDS PER GALLON OF JP-4 FUEL.
- ⚠ OVERLOAD CONFIGURATION

(A) USABLE FUEL			(B) USABLE FUEL	
POUNDS	GALLONS		POUNDS	GALLONS
1944	299	NO. 1 WING TANKS	1944	299
2021	311	NO. 2 WING TANKS	2067	318
2756	424	NO. 3 WING TANKS	2782	428
1365 ⚠	210 ⚠	TRANSFER TANKS	1443	222
1560	240	FUSELAGE TANK	1144 ⚠	176 ⚠
195	30	LINES	45	7
2951	454	EXTERNAL TANKS	2951	454
8476	1304	TOTAL USABLE FUEL NORMAL CONFIGURATION	8281	1274
9841	1514	TOTAL USABLE FUEL OVERLOAD CONFIGURATION	9425	1450
12,792	1968	TOTAL USABLE FUEL WITH EXTERNAL TANKS	12,376	1904
8717	1341	FULLY SERVICED NORMAL CONFIGURATION	8473	1303
10,082	1551	FULLY SERVICED OVERLOAD CONFIGURATION	9594	1476
13,072	2011	FULLY SERVICED WITH EXTERNAL TANKS	12,584	1936

The fuel cells in the 106's wet wing are clearly seen in this schematic.   (USAF Photo)

Details of the "Six's" vertical tail and flaps flow beautifully with aerodynamic cleanness. (USAF Photo)

Details of one of the two rear wheels of the F-106 landing gear. The wheels appear very small compared to the size of the aircraft.

(Dale Witt Photo)

All fuel and armament are carried internally on the "Six" to preserve the clean, trim lines for maximum top speed and aerodynamic efficiency. However, for ferry flights beyond the normal range of the aircraft, the "Six" can be equipped with dropable auxiliary fuel tanks.

Actually, when one gets right down to it, the "Six" was the last aircraft designed to carry all its ordnance internally. A 106 pilot commented that he thought "People had forgotten what a nice clean aircraft flies like." What he was referring to is the recent trend of designing aircraft with multiple hard points on the wings and fuselage for the mounting of external ordnance. The "Six" was designed from the first day as an interceptor. It does its job in a super efficient manner and a large part of that efficiency is derived from its ultra-clean aerodynamic shape. And in this day of hang-on ordnance, for the "Six" to survive clean for almost 20 years is amazing.

This is not to say that the "Six" has not been modified and improved. But a majority of the improvements have come internally. F-106A's are now cavorting with clear canopies, resolving the visibility problems that long plagued the "Six." Many other changes have been made, or are planned, to stretch the lifetime of this sleek bird. (These changes will be discussed later in this book.)

Like all true deltas, the F-106 does not have a horizontal stabilizer. Control is achieved through movable "elevon" surfaces in the trailing edge of the wing. These surfaces function both as elevators and ailerons. Clamshell dive brakes are located at the base of the vertical stabilizer.

Among the pilots who fly her, the "Six" is regarded as a "forgiving" aircraft. Possessing exceptional stability in high-speed flight, as well as ease of handling at low speed, the "Six's" delta wing design makes it a superb flying machine in all flight regimes. It should also be mentioned here that the B version of the "Six," manufactured in a two-place tandem configuration, has virtually the same appearance and performance as its single-seat sister ship, and in addition has a trainer capability.

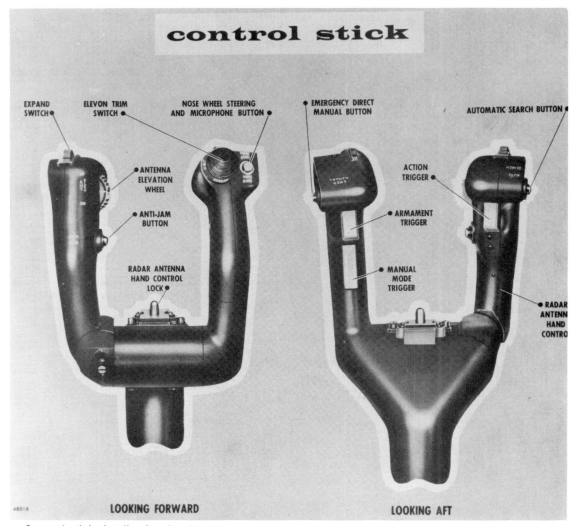

control stick

EXPAND SWITCH

ELEVON TRIM SWITCH

NOSE WHEEL STEERING AND MICROPHONE BUTTON

EMERGENCY DIRECT MANUAL BUTTON

AUTOMATIC SEARCH BUTTON

ANTENNA ELEVATION WHEEL

ACTION TRIGGER

ANTI-JAM BUTTON

ARMAMENT TRIGGER

RADAR ANTENNA HAND CONTROL LOCK

MANUAL MODE TRIGGER

RADAR ANTENN HAND CONTRO

LOOKING FORWARD

LOOKING AFT

Control stick details showing location of multitude of controls.

(USAF Photo)

The cockpit shown in this photo illustrates the early design. F-106's have now been equipped with the completely clear model. (Dale Witt Photo)

Instrument panel of one of the F-106A test aircraft.

(Convair Photo)

F-106A early version left-hand console.

41

cockpit general arrangement (typical)

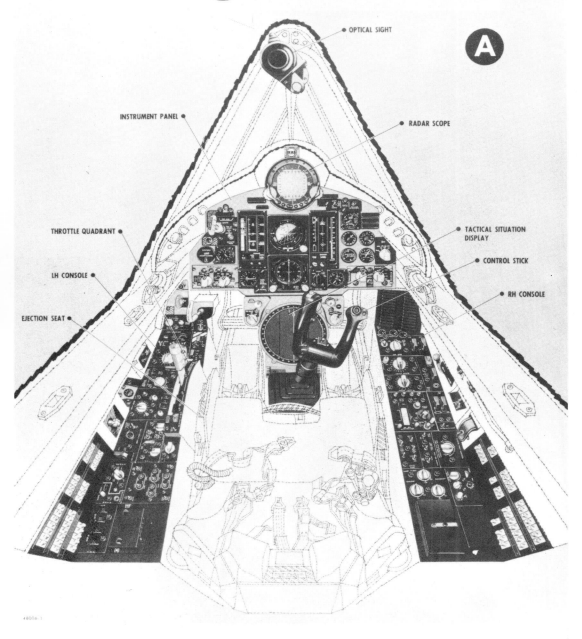

Artist's concept of F-106 cockpit showing compactness.

(USAF Photo)

One of the initial innovations incorporated into the "Six" was its so-called "supersonic seat." As the name indicates, it allowed the 106 pilot to eject safely at supersonic speeds. The seat was developed as a result of an industry-wide meeting in 1956, a time period when a multitude of high performance aircraft were developed. This seat did not prove as successful as hoped. In fact, it was replaced in the F-106 fleet in the 1964-67 time frame by the "Weber" seat.

The initial configuration seat folded the pilot's knees against his chest and snugged his arms between his thighs and stomach before launching him. Following canopy jettison, two rearward mounted booms extended stabilizing the chair. The parachute system was armed when the seat was launched from the aircraft, but the system was not activated until it was below 15,000 feet.

Extensive testing of the initial ejection system was conducted at Edwards Air Force Base and Holloman Air Force Base. In one sled test at Edwards, a package of cigarettes remained in the dummy's breast pocket throughout a test run at Mach 1.4 at sea level, simulating Mach 2.5 at 30,000 feet. At Holloman, 35 human sled runs were successfully conducted. These verified that accelerations imposed during ejections up to 900 knots were within human endurance.

The "Six" was the first aircraft to have a digital computer built into the fire control system. Through the years the system has been updated with solid-state electronics. But initially the system was probably beyond the state-of-the-art which hurt in terms of reliability.

The "Six's" MA-1 navigation and armament control system is one of the most complex electronic systems in military aviation—initially consisting of some 200 black boxes, thousands of tiny electronic parts and nearly eight miles of wiring. Yet this system can check itself out before a flight in just five minutes.

Hughes developed a semi-automatic procedure for "go or no-go" preflight testing called the short-system ground check. The check is performed with a seven position selector switch on a cockpit control box inside the F-106 which sends test signals to the MA-1 major subsystems and to the MA-1 digital computer for digestion.

A self-test program is magnetically recorded and stored on tape in the computer's memory system. The computer, the brain of the MA-1 control, knows what the results of the test should be and monitors the test signals between the cockpit and the subsystems. Isolation of malfunctions is based on electronic signals received inside the cockpit from a mobile automatic radiating tester (MART) and self-test units built into each of the five major MA-1 subsystems—communications, navigation and landing; automatic flight control, radar, armament and the computer.

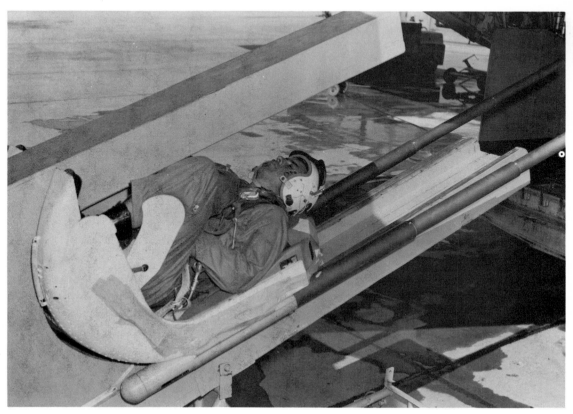

Jumpers in preparation for F-106 seat "live jumps."   *(U.S. Navy Photo)*

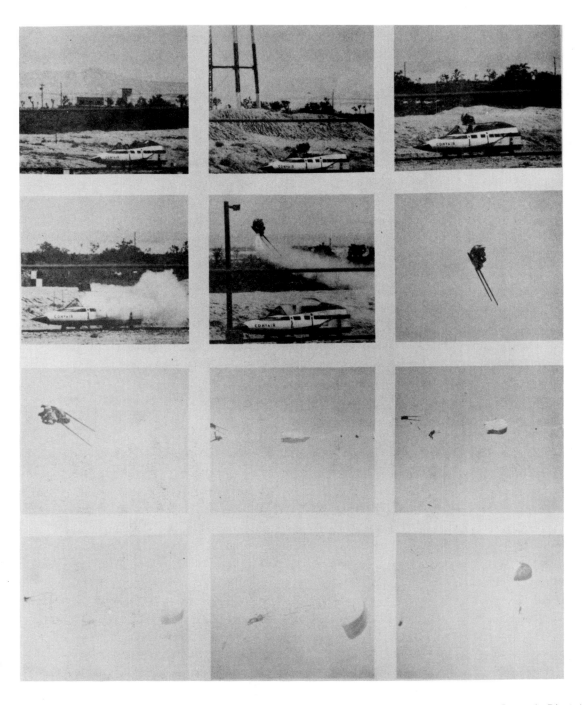

Early sled testing at Holloman Air Force Base at the F-106 supersonic ejection seat. (Convair Photo)

F-106B test aircraft for ejection seat testing at Holloman Air Force Base.

(USAF Photo)

An F-106A is shown with the 200 black boxes which comprise the MA-1 armament control system. At right is the Douglas Air-2A Genie Nuclear Rocket with AIM-4F and AIM-4G Super Falcon missiles on the left.

(Hughes Photo)

"People have forgotten what a clean aircraft flies like," said an F-106 pilot. All ordnance is carried internally within this weapons bay.
(Dale Witt Photo)

The MART, an unmanned mobile radio unit mounted on a four wheel undercarriage, is used to check the communications, navigation and landing subsystem. On a command signal from an aircraft, the MART transmits a 52 second test program to the cockpit through the short system ground check control panel. The MART is positioned on the flight line within a one mile transmitting radius of all aircraft. Only one MA-1 control system may be tested at a time. If a technician dials the MART and the unit is in use, he gets a "busy signal," and he has to call again.

The other subsystems check themselves for malfunctions as they receive the test signals from the cockpit control panel. The computer verifies the results. Signals are transmitted to the cockpit, and if the systems are operating properly, the signals activate a set of lamps on the panel. Turning the selector switch to the next position automatically switches off the lamps and starts another sequence of tests.

There is a lamp on the cockpit panel for each major subsystem. All the lamps must light in each of the seven tests before the technician releases the aircraft to the pilot for a mission. If one of the lamps fails to light, there is a malfunction somewhere in the corresponding subsystem. With a quick inspection, the technician can find the black box causing the trouble and replace it with an operating unit.

Getting down to the business end of the "Six," it is necessary to crawl under the wing and open the flush-mounted missile bay. Here, all the weapons are

carried internally with a full weapons load consisting of two AIM-4F and two AIM-4G Super Falcons, along with a single nuclear-tipped AIR-2A Genie rocket.

The AIM-4F Super Falcon was introduced in 1960. It has an improved radar guidance system with greater accuracy and increased immunity to enemy decoy signals and countermeasures which are applied by invading bombers to confuse the radar systems of defending interceptors and their missiles. The AIM-4F is powered by a new solid fuel, two-level thrust rocket engine that produces high launching thrust followed by a lower level thrust to sustain missile velocity. The engine operates with high reliability under the great extremes of temperature encountered by the F-106. The forward portion of the AIM-4F is covered by a white, moisture sealing sleeve. The missile is 86 inches long, 6.6 inches in diameter, has a wing span of 24 inches and weighs about 150 pounds. More than 3,400 have been built.

The AIM-4G Super Falcon is the infrared seeking counterpart of the AIM-4F. It is equipped with a newly developed infrared detector that enables it to lock on to smaller targets at greater ranges than early infrared missiles. The AIM-4G is carried in mixed loads with the AIM-4F in the F-106. Except for the infrared nose cone, both missiles have the same external configuration. The AIM-4G is 81 inches long, 6.6 inches in diameter, has a wing span of 24 inches and weighs about 145 pounds. About 2,700 units have been built.

47

F-106 firepower in the form of Genie rocket and Falcon missiles.   (USAF Photo)

Low angle view of AIR-2A Genie on lifting arm underneath F-106.   (USAF Photo)

Maintenance is an always-there job. These "Six's" belong to the 318th FIS at McChord AFB, Washington. *(USAF Photo)*

Sand bag revetments are in evidence as F-106's serve their time in Korea.   *(USAF Photo)*

A 318th FIS pilot talks to a ground crewman before taking to the air.   *(USAF Photo)*

Photo illustrates testing of new weapon concept for the "Six."   *(USAF Photo)*

An artist's concept of the proposed (but never to be) F-106X configuration.   *(Convair Photo)*

Tall tail against the sky.   *(Dale Witt Photo)*

Looking more like a rocket, the wide-angle camera shot catches "Six's" beauty.
(Dale Witt Photo)

F-106 intake detail.   *(Dale Witt Photo)*

The Air Force Museum's F-106 is shown with some of her contemporaries—an F-86B, an F-101, and an F-100 among others. *(Dale Witt Photo)*

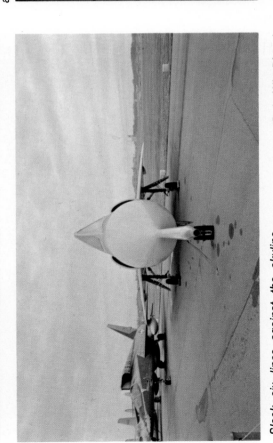
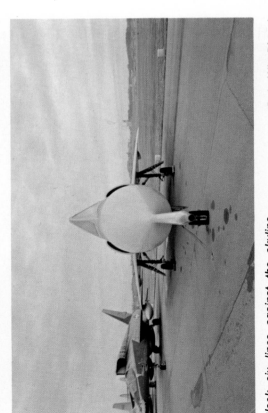

A sign of the 1960's. An F-102 moves in for an inspection of a Soviet turbo-prop Bear Bomber. *(USAF Photo)*

(Dale Witt Photo)

Distorted shot of AF Museum's F-106.

(Dale Witt Photo)

Sleek six lines against the skyline.

A pair of F-106's cruising above the cloud layer.   *(USAF Photo)*

F-106 Unit patches and "Dart" shoulder patches.   *(Robert Shenberger Photo)*

ADC's 11th FIS (since deactivated) performs maintenance on a "Six" in an a-
lert hangar at Duluth International Airport, Minnesota, in 1967. (USAF Photo)

(USAF Photo)

An F-102 pilot watches his wingman peel off.

USAF Personnel remove the container tops of ADC Falcon Missiles while an F-106 waits in the background to be loaded at Richards-Gebaur AFB, Missouri.   *(USAF Photo)*

This F-106 receives maintenance with the 11th FIS at Duluth International Airport in 1967.
(USAF Photo)

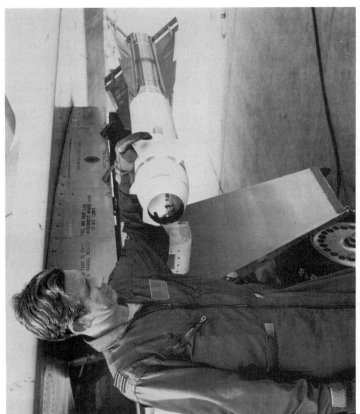

Squadron Leader Keith Hawes stands next to an AIM-4 Falcon missile mounted under the wing of an F-106 of the 4750th Test Squadron.

(USAF Photo)

Major John Fowler stands next to two Falcon missiles at Holloman AFB, New Mexico, in 1959.

(Convair Photo)

FLIGHT MANUAL
USAF SERIES **F-106A** & **F-106B** AIRCRAFT

AF SERIAL NO. 57-2508 AND ON

THIS PUBLICATION REPLACES T.O. 1F-106A-1 DATED 6 JANUARY 1961, AND SAFETY OF FLIGHT SUPPLEMENTS -19 THROUGH -29, -31 THROUGH -36, -38 THROUGH -40; AND T.O. 1F-106B-1 DATED 20 JANUARY 1961, AND SAFETY OF FLIGHT SUPPLEMENTS -17 THROUGH -27, -29 THROUGH -34, -36 THROUGH -38. SEE WEEKLY INDEX, T.O. 0-1-1A, FOR CURRENT STATUS OF SAFETY OF FLIGHT SUPPLEMENTS.

THIS PUBLICATION IS INCOMPLETE WITHOUT CONFIDENTIAL SUPPLEMENT, T.O. 1F-106A-1A.

PUBLISHED UNDER AUTHORITY OF THE SECRETARY OF THE AIR FORCE.

COMMANDERS ARE RESPONSIBLE FOR BRINGING THIS PUBLICATION TO THE ATTENTION OF ALL AIR FORCE PERSONNEL CLEARED FOR OPERATION OF SUBJECT AIRCRAFT.

(USAF Photo)

Chapter III

Flying the Bird

(NOTE: Portions of this Chapter appeared as an article entitled "Flying the Six" by Captain Donald D. Carson in the October 1973 Air Force Magazine. Their permission for its inclusion into this book is appreciated.)

How does an aircraft perform after ove sixteen years of hard operational use? The men who fly the "Six" think she has improved with age. Many say the bird is one of the truly great airframe designs of modern aviation. In pilot circles it is affectionately coined "The Cadillac of the Fighters." The "Six" can perform its mission far better today than it could when introduced in 1959, because the systems have been continually refined.

The physical beauty of the F-106 is immediately apparent. Its sleek fuselage and its tall, sweptback tail give an indication of the aircraft's great speed. The F-106 established several altitude records, and, in 1959, set a world's official speed record of 1,-525.95 mph, which is impressive even today. The F-106 has been the first-line interceptor of ADC and NORAD since 1959.

"To give you an idea of what it is like to fly the F-106, let me* take you along a typical training mission illustrating its interceptor capability.

"Our walk-around inspection starts with the lance-like pitot tube at the very front of the aircraft. This provides an air-pressure input for the central air data computer (CADC), which in turn provides accurate airspeed and altitude information to the flight instruments and main aircraft computer.

"Passing under the wing, we continue the inspection, stopping to open the missile bay to inspect our weapons load. Today, we'll be firing live AIM-4 Falcon missiles on the air-to-air range over the Gulf of Mexico, near Tyndall Air Force Base, Florida. A full weapons load consists of two IR and two radar-

---

* *'Me' being Captain Donald D. Carson (former F-106 pilot).*

Major John Mantei, former F-106 test project officer, explains the characteristics of the Delta Dart to the author.
                                                                (Dale Witt Photo)

Wheels up! An F-106 pours the power to it during a takeoff.   (USAF Photo)

guided missiles and an AIR-2A Genie rocket. Today's firing load is two AIM-4F radar missiles. The three types of air-to-air weapons give the F-106 an excellent capability against either manned bombers or maneuvering fighters at both high and low altitudes. All armament is carried internally.

"Our exterior inspection complete, we climb the ladder into the cockpit. Our first check is the vertical tape instruments, which are used instead of conventional round gauges. Once you've flown a "taped" bird, you are forever spoiled. Tapes present all necessary information in such a clear manner that it is almost impossible to misread altitude or airspeed. The "Six" has a well laid-out cockpit although not as large as an F-101 or F-4. The F-106 general design philosophy was a 'heads-down' concept where the pilot is flying mostly from displays in the cockpit without looking out. A good part of his time is spent analyzing the radar scope. The bird was designed to be flown by the autopilot where the pilot would be controlling the throttle and selecting the ordnance.

"Centered above the aircraft instruments is a special 'daylight' radarscope. The scope background is a bright green with white target returns, easily visible in broad daylight. Older scopes needed a hood to shade them, or else the pilot had to lean forward to see the scope displays.

"A unique feature of the 'Six' is the 'annunciator' for the armament, computer, and navigation systems. A small, round indicator window tells the status of each system. There is never any doubt as to whether they are operating or not.

"On the lower pedestal, between my feet, is one of the most remarkable pieces of navigation equipment ever put into a fighter—the Tactical Situation Display (TSD). It resembles a TV screen and shows a map corresponding to the TACAN navigation station I've selected. A triangle, called the interceptor symbol, which represents my aircraft, is positioned over this map at our exact location. The advantages of this versatile system become evident especially during a night weather penetration.

Two F-106's are framed by some weeds on a low-level run.   *(USAF Photo)*

"After we're strapped in, I depress the engine ignition button and move the throttle outboard and then back in to fire the starter motor and provide ignition. The engine can be started without external power by using internally stored high-pressure air and the aircraft battery. This enables the F-106 to operate from dispersed airfields with a minimum of support.

"Once started, I turn on the single MA-1 fire-control power switch, which operates all of the weapons, radar, computer, navigation and communication equipment. I dial in a grid reference setting to tell my computer the location and aircraft heading. The aircraft computer has tremendous capabilities, and one of them is dead-reckoning navigation. Once the grid reference setting has been inserted, I can fly to any predetermined fix on my TSD without receiving information from a TACAN station or any other type of navigation aid.

"I close the canopy and taxi to the runway. Everything looks good, so I 'hack' the clock, release the brakes, and put the throttle in afterburner. **Suddenly everything gets quiet for a moment. Then the J75 really comes to life.** I"m jolted forward by a solid kick in the back and a loud bang as I get the 'hard light' so characteristic of the J75 engine. This is the same engine found in the F-105, **making the "Thud" and "Six" the two most powerful single-engine aircraft in the world.** The J75 puts out 24,500 pounds of thrust in full afterburner (26,500

for the F-105 during a water injection takeoff). The hard light is even more apparent than in the F-105, as the "Six" is several tons lighter.

"Acceleration is extremely rapid. I ease back on the stick at 135 knots to raise the nosewheel off the runway. Holding this takeoff attitude, the aircraft flies off the runway at 184 knots. At 250 knots, I come out of afterburner long before crossing the end of the runway. Moving almost 42,000 pounds from a standing start to more than 250 knots in about 7,000 feet is quite impressive. The F-106 is a thrill to fly, and the novelty never wears off. I accelerate out to 400 knots and begin to climb at a steeper rate, maintaining this speed until reaching Mach .93, which I hold to level off. I kick my rudders to fishtail the aircraft—a signal to my wingmen that I want them to move out into route formation.

"After contacting the Ground-Controlled Intercept (GCI) director who will control the mission, I separate my flight. Each aircraft begins to follow the 'Data Link' commands sent by the intercept director through the ground control center computer. Under Data Link direction, the computer at the Semi-Automatic Ground Environment (SAGE) or Backup Interceptor Control (BUIC) center transmits information to each aircraft. The MA-1 aircraft computer displays data as heading, airspeed, and altitude commands. I also receive target heading, speed, altitude, range, and bearing information.

Pilots who have flown the F-106 say that it is probably one, if not the smoothest, flying machine ever built. (USAF Photo)

Like two birds in flight, two Delta Darts are silhouetted against the setting sun.   *(USAF Photo)*

F-106 night takeoff.   *(USAF Photo)*

"Once I've checked in with my intercept director, giving my armament safety check, the remainder of the intercept can be conducted without either of us saying a word. I receive all commands on my 'tapes' in the form of white markers that appear over the speed, altitude, and heading I'm to fly. There is also information displayed on the Tactical Situation Display (TSD), which depicts the entire intercept on my display. I can see my position in relation to that of the target, and the type of intercept I'll be conducting. Today, for range safety, I'll call my contact with the target and get verbal clearance to fire from my GCI controller.

"When the target-marker indicator moves up on the altitude tape, and I begin to receive target range, I know I've been committed against a specific target. At this time I arm my missiles.

"I search the sector of my radarscope that corresponds to the target bearing and distance being sent by the Data Link. I position my radar antenna elevation to search the altitude at which my target is flying. Today, I'll be directed to make a 10,000-foot front 'snap-up' attack against a Firebee drone target flying at 40,000 feet.

"On turning toward the drone, which is now thirty miles ahead, coming directly at me. I select afterburner to gain speed for the snap-up. The snap-up maneuver is used against targets at very high altitudes. This drone will not be above 45,000, but I'll still use a snap-up since it is a more demanding intercept and provides very realistic training. The afterburner quickly pushes me through the transonic area into supersonic flight. There is no difference in the feel of the aircraft as it goes supersonic. Your only indication is a slight movement in the altitude tape, which quickly settles back down to normal.

"I spot my target five degrees left at the top of my scope and call a 'contact.' Grasping the left half of the 'split stick,' which controls both the aircraft and the radar system. I'm positioning the antenna beam and 'range gate' over the radar return. The radar locks on. 'Red Lead . . . Judy,' I call to the GCI controller to indicate I'm assuming full control of the intercept.

"The MA-1 computer now takes over and computes the intercept steering geometry. I can either select the 'auto-attack' mode, which will take the computer inputs and steer me to the target, or fly it manually. The autopilot doesn't need the practice! I'm turning to center the steering dot depicted on the radar attack displays. The target is moving rapidly down the scope. I'm selecting the expanded sixteen-mile radarscope display, which gives more precise information.

Four six's flying an almost perfect formation over Maryland in 1970.   (USAF Photo)

Straight and level. F-106 on alert duty.

(USAF Photo)

(USAF Photo)

Diamond formation of four Delta Darts in perfect formation.

66

Two "Six's" from the 318th FIG appear to be sitting on each other in this interesting picture.

(USAF Photo)

"At approximately fourteen miles, the scope tells me it's time to begin the snap-up. I'm smoothly pulling the nose above the horizon into a steep climb as the outer radar range circle on the radarscope begins to shrink. When this circle shrinks to the same size as the smaller steering circle, the missiles will fire. A steering dot and another smaller circle on the scope provide directional information. The aircraft is turned to put the 'dot in the hole,' thus positioning the aircraft for an accurate missile launch.

"Looking up, I see the drone dead ahead and well above me. Squeeze the trigger! Wait for the computer to fire the missiles at the correct moment! The steering dot is 'pegged' directly in the center of the steering circle. When the fire signal appears on the scope, there is a loud rush of air as the weapons bay doors rapidly slam open.

"Now a roar as two Hughes Falcon missiles accelerate away from me as if I were sitting still. They're heading toward the drone with a closure rate almost three times the speed of sound. It's a hit!

"Back in the airfield traffic pattern, I'm reminded of one disadvantage of the delta wing—the absence of flaps. This causes the "Six" to have relatively high final approach and landing speeds. A normal weight final approach (2,000 pounds of fuel remain-

ing) is flown at 181 knots, with touchdown at 149 knots. Landing speeds can exceed 200 knots on final with a heavy fuel load on board. However, the drag chute and high drag generated by the delta wing during aerodynamic braking enable you to stop the F-106 in very short distances. Aerodynamic braking is accomplished by slowly raising the nose of the aircraft—up to a maximum of seventeen degrees—once your main landing gear has touched the runway. Any more than 17 degrees and you drag your tail. It gives you the feeling that you're going to topple over backwards. The slowdown is also aided by speed brakes which are operated from the throttle."

Another 106 pilot commented that the landing gears are fairly small for the weight and size of the aircraft coming down. Like most aircraft landings, if a tire were to blow there is a good chance the aircraft would be destroyed.

The high angle-of-attack landing approach presents a real problem to the pilot should he be unable to get the landing gear down. The prescribed policy under such circumstances is to "punch out." Were you to land wheels up you would first drag the aft portion of the fuselage with the nose being slammed to the pavement with a great force.

Two "Six's" carrying their "supersonic" ferry tanks fly in close formation.

(USAF Photo)

The underbelly of this F-106 appears to be aflame as a Genie rocket streaks toward its target. (USAF Photo)

The weapons bay of this F-106 is still open after it has discharged its ordnance. (USAF Photo)

An F-106A, from the 84th FIS, makes a perfect landing at Elmendorf AFB, Alaska in 1969. (USAF Photo)

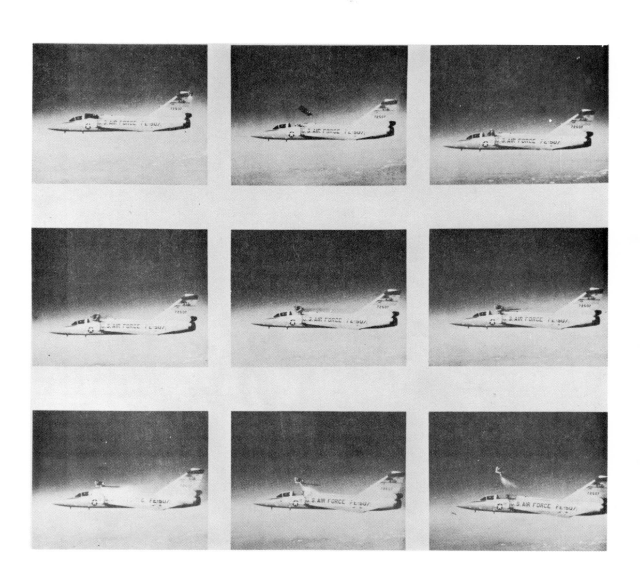

The series of photographs above illustrates an actual in-flight ejection from Aircraft F-106B S/N 57-2507 on September 2, 1960, at Holloman AFB, N.M. The aircraft was flying at 50,000 feet altitude and Mach 0.95 when the seat was ejected from the rear cockpit. All components of the recovery system functioned satisfactorily and the dummy was recovered. *(USAF Photo)*

An interesting array of F-106 fighter compatriots. Shown are counterclockwise top-to-bottom F-105,
F-101, F-102, F-100 and F-104.　　　　　　　　　　　　　　　　　*(USAF Photo)*

The F-106B has few changes differentiating it from the single-seat version.

"Six's" of the 318th FIS at Elmendorf AFB, Alaska in 1963. These "Six's" were the first 106's to be deployed to Alaska.
(USAF Photo)

Chapter IV

Operational Service

The "Six" is a unique bird in many ways. First, it is the only operational pure delta wing aircraft in the active USAF inventory. It may well be the last! Secondly, unlike many of its fighter brothers (i.e., the F-4, F-5, A-4 and others) the 106 has never seen service with the Air Force of any other country. And finally, even though she has been on active duty for 16 years, she has never been forced to fire a shot in anger.

transition as the Air National Guard is assuming "Six's" in large numbers in its role with the Aerospace Defense Command (ADC).

Mustered for full-time operations more than two decades ago, ADC is still on around-the-clock alert status. ADC has changed over the years from a command based primarily in the U.S. and concerned with defense against the bomber to one spread worldwide to warn of any attack through air or space. Today, as the major component of the U.S.-Canadian North American Air Defense Command

(NORAD), it trains and operates units equipped for the jobs of air defense, aerospace surveillance and early warning.

The units ADC trains to combat-ready status are made available to NORAD to perform the missions of aerospace surveillance, advance warning and assessment of missile attack, peacetime air sovereignty and limited bomber defense in times of crisis.

The Aerospace Defense Command had a seesaw beginning. First established in 1946, it was consolidated two years later, along with the Tactical Air Command, under the Continental Air Command. By 1950, Continental Air Command had assumed full charge of air defense and ADC was abolished. With a new build-up of air defenses and a change in Air Force command structure, ADC was re-established in 1951.

As the Soviet bomber threat developed, so did ADC. The command reached its peak strength in

The nose of this ADC F-106 frames a maintenance airman during the early hours of the morning.
(USAF Photo)

Two "Six's" of the now-deactivated 71st FIS are framed by the nose section of another. Photo was taken in 1967 at Richards-Gebaur AFB, Missouri.   (USAF Photo)

1957 with 69 fighter interceptor squadrons, 1,500 aircraft, and more than 100,000 people committed to its air defense mission. Defenses against the bomber threat grew and improved during the 1950's.

The Distant Early Warning (DEW) Line was strung along the Arctic Circle to watch for signs of bombers approaching North America over polar routes. Other ADC radars were strategically positioned along the U.S. coasts and in the interior of the country. Coverage was extended off shore by early warning aircraft and radar platforms on steel stilts embedded in the Atlantic ocean floor, commonly called, "Texas Towers." BOMARC unmanned ground-to-air interceptor missiles complemented the fighter force.

A semi-automatic command and control system was put into operation at centers across the nation. Electronic computers were installed in these centers to quickly process masses of radar data and compute information needed to direct air defense interceptor weapons. While the 1950's witnessed the building of a vast system to defend against bomber attack, the end of the decade saw the coming of a new and more potent threat—the intercontinental ballistic missile. Air defenses had to be restructured to provide the quicker warning needed for a missile attack.

Today the command has a perimeter system of aircraft-detection radar sites along U.S. borders, as opposed to the overlapping interior and exterior coverage of the 1950's. The last of the BOMARC missile units was inactivated in 1972. A backup interceptor control system deployed to augment the primary command and control posts was inactivated in 1974 with the exception of one site in Florida. Radar carrying aircraft that once extended around-the-clock early warning off both coasts have been cut for continental U.S. use to a force that now makes random patrols along the west coast and off southern Florida.

ADC still maintains a reduced force of fighter interceptors. They police national air space in peacetime. In event of war, they would be augmented by fighters of other Air Force commands, the U.S. Navy and Marines for use by NORAD in combating a bomber attack.

ADC has six squadrons of "Six's". Active duty units equipped with the "Six" (as of early 1976) were the 5th Fighter Interceptor Squadron (FIS), Minot AFB, North Dakota; the 48th FIS, Langley AFB, Virginia; the 49th FIS, Griffis Air Force Base, New York; the 84th FIS, Castle Air Force Base, California; the 87th FIS, K. I. Sawyer Air Force Base, Michigan; and the 318th FIS, McChord Air Force Base, Washington.

3/4 rear view of an F-106 waiting on the ramp to be loaded with Falcons and a Genie at Richards-Gebaur AFB, Missouri in 1967. (USAF Photo)

A "Six" pilot of the now deactivated 539th FIS is signaled into the parking area by his crew chief at McGuire AFB, New Jersey, in 1960. (USAF Photo)

95th FIS "Six's" with drag chutes extended are seen through concertina as they taxi in at Osan Air Base, Korea. The aircraft replaced F-106's of the 94th FIS, which returned to the states.

(USAF Photo)

The coke-bottle fuselage of the F-106 is clearly evident in this photo as 95th FIS maintenance men keep her ready to go.
(USAF Photo)

Rear view of 95th FIS "Six's" during a snow in Korea. (USAF Photo)

A 95th FIS 106 with drag chute extended is seen behind concertina as it taxis in at Osan Air Base, Korea.
(USAF Photo)

80

An interesting, and significant, highlight should be mentioned about the latter, the 318th FIS. For the first time in history, United States Air Force F-106's using inflight refueling flew to a critical overseas area along with tactical air units. "Six's" assigned to the 318th Fighter Interceptor Squadron were flown to Osan AB, Korea as part of the Air Force build-up triggered by the North Korean Pueblo crisis. Armed with Falcon air-to-air missiles, the 1,400 mph Convair-built fighters arrived at their destination ready to assume a combat ready posture. The planes were immediately placed on alert to provide air defense for other forces in the area. Other F-106 units would later follow the 318th to Korea.

The deployment was a joint operation with SAC KC-135 tankers providing the aerial refueling. The previously modified for air-to-air refueling fighters carried the then-new external 360 gallon fuel tanks, and had the capability of receiving fuel at an average rate of 2,000 pounds (over 300 gallons) per minute from the tankers.

The Air National Guard currently (1977) has six F-106 units and a total of 90 F-106 aircraft. The first unit to receive the F-106 was the 120th Fighter Interceptor Group, Great Falls, Montana, which began conversion in April 1972. The next three units, which received the F-106 during the second half of FY 1973, were: 102nd FIG, Otis ANGB,

In 1967, the F-106 fleet was modified for mid-air refueling. Here a "Six" gets a load of JP-4 from a Boeing/USAF KC-135. *(USAF Photo)*

Three F-106A's of the Montana Air National Guard fulfill their commitment to the ADC Mission.
(USAF Photo)

The familiar "Big Sky Country" insignia adorns the vertical fin of the Montana Air Guard's "Six's."
(USAF Photo)

Massachusetts; 177th FIG, Atlantic City, New Jersey; 191st FIG, Selfridge ANGB, Michigan. Two additional units, the 125th FIG, Jacksonville, Florida, and the 144th FIG, Fresno, California, began converting to the F-106 in July 1974, and are now operational.

An interesting note which typifies the "can-do" attitude of Air National Guardsmen is that 59 days after receiving 50% of their new aircraft, the Montana unit placed two F-106's on active air defense alert. Guidelines established by ADC had called for this transition to air defense alert to be accomplished in approximately 18 months.

The Aerospace Defense Command is a deterrent to direct attack. The command tells the enemy he cannot count on surprising us—and that an indeterminate portion of his attacking forces would never reach their targets in this country. Where does today's aerospace defense team acquire these essential skills necessary to detect . . . intercept . . . identify and destroy any hostile fighter and bomber aircraft and thus provide the vital deterrent?

Charged with an awesome responsibility is the Air Defense Weapons Center at Tyndall Air Force Base, Florida. This is where expertise in air defense is expected as part of everyday living. It's at Tyndall where ADC F-106 fighter-interceptor pilots undergo an annual weapons firing program . . . where pilots get advanced training in their 106's . . . where pilots learn the latest tactics . . . and where tests are conducted by the Aerospace Defense Command to make sure new equipment and tactics fit the defense mission.

At least once a year, every Aerospace Defense Command F-106 pilot comes to Tyndall to pit his skills against the BQM-34A Firebee drone, a radio-controlled target that effectively simulates an "invader" aircraft. The drone, operated by remote control, simulates an enemy aircraft invading American airspace.

Side view of four ADC F-106's of the 48th Fighter Interceptor Squadron, Langley AFB, Virginia, in a 1970 flight over the Florida coast.   *(USAF Photo)*

83

Side view of an ADC 438th FIS (now disbanded) F-106 taxiing out as seen from inside the mobile control unit at Kincheloe AFB, Michigan, in 1967. *(USAF Photo)*

F-106A being attached with a tow-bar for towing to a hangar for maintenance. (USAF Photo)

Periodically, "Six's" receive depot maintenance performed by the Air Force Logistics Command. (USAF Photo)

Impressive view of 106 tail sections at Dover Air Force Base, Delaware. The aircraft were part of the now deactivated 95th FIS. (USAF Photo)

F-106 fighter squadrons participate in the program as a unit once each year. Objectives of this program are designed to determine overall ADC interceptor systems capabilities and effectiveness. Each deploying unit at the weapons center is assigned different test conditions to satisfy the overall command objectives.

Although the weapons testing involves the entire squadron, only a limited number of aircraft deploy to the Weapons Center at any one time. As each pilot completes his mission, he returns to his home base and another pilot takes his place. The rotation continues until each aircraft in the squadron has been flown and its weapons systems qualified. Air National Guard 106 units undergo the same rigid training program at Tyndall on a yearly basis.

Another important role in the air defense mission is performed by the 2nd Fighter Interceptor Training Squadron. This squadron conducts advanced flying training programs for the "Six." A class of student pilots spends its first two weeks studying academics before they even get into a cockpit. After a solo flight and training in a flight simulator, students are introduced to the sophisticated techniques of modern aerial combat. They practice every conceivable type of intercept; the tactics of high, medium and low altitude intercepts during day and night, along with infrared and radar attacks.

Along with the training missions for pilots and weapons controllers, the Air Defense Weapons Center at Tyndall AFB has the task of training experienced officers and instructors. This is accomplished at the 62nd Fighter Interceptor Training Squadron's Interceptor Weapons School (IWS). IWS also serves the overall mission of the United States Air Force in many other ways. IWS participates in weapons tactics development for both current and future air-to-air weapons systems. IWS develops and conducts a formal aerial combat tactics training program for ADC F-106 pilots. This program prepares the pilot to operate his weapon system at maximum effectiveness in the fighter versus fighter air defense role, where the threat is a high speed, highly maneuverable, enemy aircraft. This environment could be expected while protecting or attacking a tactical strike force in a radar-controlled environment.

The school participates with other major commands and services in studies of weapons, tactics and operating procedures to be employed wherever the role of air defense exists. These joint developments are an example of the way the various commands utilize their knowledge and specialized capability to provide USAF fighter aircrews with the most current information and techniques in air-to-air weapons delivery. These programs are essential to keep F-106 pilots trained to a keen edge, and provide a basis for the USAF missions and training requirements.

An ADC F-106, of the 48th Fighter Interceptor Squadron, stands alert in heavy fog which has rolled in from the Chesapeake Bay.   *(USAF Photo)*

"Six's" of the 84th Fighter Interceptor Squadron line up for departure from Hamilton AFB to Castle AFB as phase-down and realignment of Hamilton AFB nears completion.   *(USAF Photo)*

A formation of Delta Darts peel off past the tail of a "Six's" high vertical fin. (USAF Photo)

Beautiful view of F-106's returning from mission. Note old-style cockpit configuration. (USAF Photo)

USAF crew chiefs Sgt. Carl E. Mross and Sgt. William Hyatt clean snow from a 95th FIS F-106 at Osan Air Force Base, Korea.
(USAF Photo)

90

Night view, taken with infrared camera, of a "Six" of the 95th FIS parked in a hard shelter at Osan.   *(USAF Photo)*

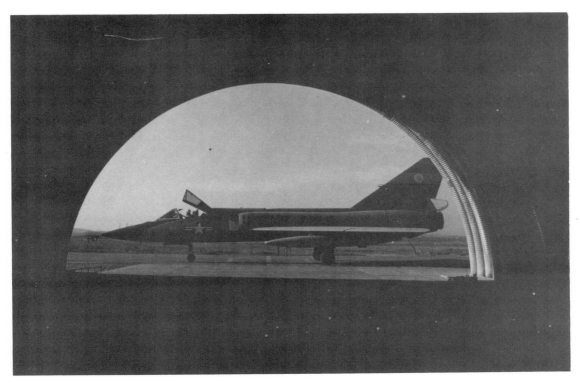

A 106 framed by a hard shelter.   *(USAF Photo)*

One of the 95th FIS 106's undergoes maintenance at Osan Air Base.

(USAF Photo)

Chapter V

The Six—
Then, Now and Tomorrow

The "Six" is one of the most remarkable aircraft ever built. It benefited greatly from the fact that it was not the first of its breed. As has been mentioned earlier, the 106 airframe was an F-102 outgrowth so it was able to benefit from 102 mistakes. It also benefited from the advanced J-75 engine which has proved to be one of the best airframe/engine matchups ever conceived. A highly stable aircraft, the "Six" was an aircraft which just begged to be improved upon.

During the 1960's and 1970's the "Six" did see many modifications and improvements, although the airframe/engine combination remained basically unchanged. F-106A's are now flying with the aforementioned new clear bubble canopy that eliminates the great visibility problem presented by the old canopy. The F-106 fleet is also getting the composite boresight modification. This is the head-up lock-on capability mentioned earlier.

There will also be greatly increased reliability built into the MA-1 fire-control system as it is updated to increase its capabilities and accuracy. Many MA-1 components have already been converted to solid-state technology, replacing the older and less reliable equipment.

The original F-106 engine accessory drive and generator system is made up of four separate and independent generators. They have now been replaced by a more-effective single multiphase F-111 generator. It has proved to be extremely reliable and will provide all F-106 electrical power, with a saving in total aircraft weight.

Probably the most significant modification since 1959 is installation of the M-61 Vulcan, 20-mm cannon "Six-Shooter" package in the missile bay of the aircraft. This system will not interfere with the Falcon missiles, which will be retained along with the gun. The only noticeable change will be a slight bulge along the centerline of the weapons bay doors where the M-61 rotating gun barrels exit the

A comparison of the old and the new. F-106 parked next to World War I SPAD.   *(USAF Photo)*

A shiny F-106B sits in front of Base Operations at the Air Defense Weapons Center at Tyndall Air Force Base, Florida. (USAF Photo)

Strange sight for a six to have external ordnance. This photo illustrates testing of new weapon concepts for the "Six".

(USAF Photo)

Close-up view of external mounting of new ordnance for the F-106.

(USAF Photo)

"Gatling Gun" test F-106 aircraft is shown with drone during tests at Eglin AFB, Florida. (USAF Photo)

Major John Mantei is shown standing next to the "Six" in which he conducted the "Gatling Gun" tests.
His "score" can be seen just above his head.
(USAF Photo)

Underneath view of Gatling Gun pad. Some F-106's will be equipped with this modification. (USAF Photo)

The Gatling Gun program was code-named "Project Six-Shooter." This photo shows the F-106 test aircraft on jacks being bore-sighted at Eglin Air Force Base, Florida.   *(USAF Photo)*

Overall view of the testing area at Eglin AFB, Florida, showing test crew at work on the modified F-106.   *(USAF Photo)*

Side view of a modified F-106 parked at Edwards AFB, California in 1969.

(USAF Photo)

fuselage. Some F-106's are programmed to get the gun.

The "Six-Shooter" package will also include the so-called "Snap-Shoot" gunsight, one of the most advanced and accurate sights ever developed. This system, specially designed for the F-106, has proved to be deadly accurate in more than a hundred test firings against drone and dart airborne tow targets.

In 1966, the F-106 was equipped with new twin tear-shaped fuel tanks which added greatly to the "Six's" range. Designed and built by Convair, the new drop tanks hold 360 gallons of fuel each which is a 130-gallon improvement over the previous type in use. The tanks are "supersonic" rated and are able to fly at any speed the aircraft itself can attain. Then one year later in 1967, the "Six" was modified to permit aerial refueling.

Recently, the F-106 fleet went through a rigorous structural integrity investigation. The "Six" had originally been designed as a 4,000 hour airframe. But the investigation showed the old bird to be holding up amazingly well and it was recertified for 8,000 hours. This could well make its lifetime stretch into the late 1980's.

Although the F-106 was built as a tactical aircraft, its high performance characteristics caused it to be used in several NASA space research programs.

During 1963, an F-106 was being used at Edwards Air Force Base to simulate portions of spacecraft reentry and landing profiles. However, the spacecraft was the Dynasoar, which was cancelled before it was born. Then in 1968, NASA Lewis Research Center began an 18-month program to flight-test advanced inlets and exhaust nozzles for SST engines. The test bed was an F-106B. And finally in late 1970, NASA announced that as a result of F-106 tests, that plug-nozzle type jet engines had a "surprising potential ability" to reduce noise without affecting aircraft performance.

During the late 1960's, a serious effort was undertaken by the Air Force to institute a substantial upgrading of the "Six" fleet. The program was coined the F-106X Program. Basically the program was a way of taking advantage of the latest technology and methods to upgrade the "Six." The plan consisted of equipping the "Six" with a new engine, new avionics and forward control canards to provide superior air-superiority capability. A doppler radar was also planned which would provide a look-down capability.

The F-106X was an alternative to the Lockheed YF-12 Mach 3 aircraft, but due to many reasons the F-106X concept was not to be and the program was discontinued. However, even into the 1970's, contractors have continued to propose 106X-type modifications to the "Six" fleet (which presently numbers at about 230 aircraft). The proposals suggest new engines with twice the thrust of the J-75 with no increase in size, bigger inlets and the aforementioned canard surfaces. With these modifications, many feel the old "Six" would be quite competitive performance-wise with the F-15 or F-16.

As late as mid-1974, an improved radar capability was being examined for the "Six." Tests had shown in 1959 that a five foot nose extension would not degrade 106 performance. This particularly modified aircraft was known as the F-106C/D. Later in the so-called F-106E/F program, new radar system packaging concepts were considered with lesser nose extensions. The status of these possible modifications at time of writing (1977) is doubtful.

The potential for improvement of the "Six" appears practically unlimited. The F-106 has been USAF's first line interceptor for 18 long years. What the future holds for it at this time is uncertain. But one thing is certain—like a fine wine, some airplanes . . . the F-106 among them . . . actually improve with age!

— — —

You're not getting older "Six"—you're getting better!!

The Air Force Museum's F-106 in a strange environment—being towed down a freeway. During October of 1970, the museum's aircraft were moved seven miles to the new location. It could well have been termed that " 'Six's' final mission."

(Photo by Major Wm. Siuru)

AERO SERIES

A detailed look at many of the world's most famous and noteworthy military aircraft. Each book contains historical commentary, selected photographic material covering all aspects of the aircraft, technical data and specifications, four pages of color drawings, plus much more. Provides an unprecedented source of material for the modeler, military enthusiast, collector and historian.

Volumes 1 thru 20 $3.00(A) each.

ISBN 0-8168-0500-8	Vol. 1	MESSERSCHMITT ME 109
ISBN 0-8168-0504-0	Vol. 2	NAKAJIMA KI-84
ISBN 0-8168-0508-3	Vol. 3	CURTISS P-40
ISBN 0-8168-0512-1	Vol. 4	HEINKEL HE 162
ISBN 0-8168-0516-4	Vol. 5	BOEING P-12, F4B
ISBN 0-8168-0520-2	Vol. 6	REPUBLIC P-47
ISBN 0-8168-0524-5	Vol. 7	KAMIKAZE
ISBN 0-8168-0528-8	Vol. 8	JUNKERS Ju 87 "Stuka"
ISBN 0-8168-0532-6	Vol. 9	DORNIER Do-335 "Pfeil"
ISBN 0-8168-0536-9	Vol. 10	SUPERMARINE SPITFIRE

ISBN 0-8168-0540-7	Vol. 11	CHANCE VOUGHT F4U "Corsair
ISBN 0-8168-0544-X	Vol. 12	HEINKEL 100, 112
ISBN 0-8168-0548-2	Vol. 13	HEINKEL 177 "Greif"
ISBN 0-8168-0552-0	Vol. 14	MESSERSCHMITT 262
ISBN 0-8168-0556-3	Vol. 15	NORTH AMERICAN P-51 "Mustang"
ISBN 0-8168-0560-1	Vol. 16	MESSERSCHMITT Bf 110
ISBN 0-8168-0564-4	Vol. 17	MESSERSCHMITT 163 "Komet"
ISBN 0-8168-0568-7	Vol. 18	FOCKE-WULF 190 "Wurger"
ISBN 0-8168-0572-5	Vol. 19	LOCKHEED P-38 "Lightning"
ISBN 0-8168-0576-8	Vol. 20	GRUMMAN F8F "Bearcat"

NEW ENLARGED 104-page SERIES ➤

ISBN 0-8168-0580-6	Vol. 21	GRUMMAN TBF/TBM "Avenger"	$3.95(A)
ISBN 0-8168-0582-2	Vol. 21	GRUMMAN TBF/TBM SUPPLEMENT	$1.95(A)
ISBN 0-8168-0584-9	Vol. 22	BOEING P-26 "Peashooter"	$3.95(A)
ISBN 0-8168-0586-5	Vol. 23	DOUGLAS TBD-1 "Devastator"	$3.95(A)
ISBN 0-8168-0588-1	Vol. 24	BOEING B-52 "Stratofortress"	$6.95(A)
ISBN 0-8168-0592-X	Vol. 25	GRUMMAN F-14 "Tomcat"	$6.95(A)
ISBN 0-8168-0596-2	Vol. 26	GENERAL DYNAMICS F-16	$6.95(A)
ISBN 0-8168-0600-4	Vol. 27	CONVAIR F-106 "Delta Dart"	$6.95(A)